color and people

color and people

THE STORY OF PIGMENTATION

Marguerite Rush Lerner, MD
Associate Clinical Professor of Dermatology
Yale University School of Medicine

Published by
Lerner Publications Company
Minneapolis, Minnesota

The author wants to thank all those who helped in the preparation of this book—especially Aaron Lerner, Clement Markert, and Joseph McGuire at Yale University; Joseph Bagnara at the University of Arizona; and Walter Quevedo at Brown University.

Copyright © 1971 by Lerner Publications Company

All rights reserved. International copyright secured. Manufactured in the United States of America. Published simultaneously in Canada by J. M. Dent & Sons Ltd., Don Mills, Ontario.

International Standard Book Number: 0-8225-0625-4
Library of Congress Catalog Card Number: 70-128800

Contents

1. A Pigment, Melanin　　　　　　　　9

2. Color and Camouflage　　　　　　15

3. Black, White, Red, Yellow　　　　28

4. Suntan-Sunburn　　　　　　　　　32

5. Skin Spots　　　　　　　　　　　　35

6. Skin, Hair, and Eyes　　　　　　　38

7. Why Melanin?　　　　　　　　　　51

　　Index　　　　　　　　　　　　　　55

Color is all around us—in the earth and oceans, in animals and plants. In order to see colors we need light to illuminate the object we are looking at. Without light from the sun or man-made sources, and without eyes to absorb the light of different wave lengths, we could not distinguish colors—or anything else by sight alone. The simplest object would have shape, texture, softness or hardness, but no visual identity. We would not know black from white.

1

A Pigment, Melanin

Was the first man on earth black or white? Nobody knows. We have no fossils of skin from prehistoric man, so we cannot tell whether our earliest human ancestors were black, brown, pink, or white. However, there is a clue to color in dinosaurs. A paleontologist found pieces of skin preserved with the fossil remains of an ichthyosaur that lived 150 million years ago. In 1956 when a zoologist examined the delicate specimens of skin with a microscope, she saw cells containing granules that looked like the pigment *melanin*.

The best way to trace the story of skin color is to look at living things rather than at extinct animals. Skin, eyes, hair, feathers, and fur have color because they contain the black pigment, melanin. It is easy to see melanin form quickly in fungi and plants. Slice a fresh mushroom

Fig. 1. The fossil remains of an ichthyosaur, a fishlike reptile that lived 150 million years ago. Some pieces of skin were found with the fossil remains of one ichthyosaur and the cells contained granules that resembled the pigment melanin.

or a raw potato and leave it out in the air. In about 10 minutes the cut surfaces are black.

How pigment forms in animals is not so simple, but it is understandable. Melanin is the chief coloring factor in fishes, frogs, reptiles, birds, and mammals. Normal amounts of melanin produce skin color ranging from pinkish tan through black. When melanin is absent the skin is milky white, and the animal is an albino.

Melanin is made in specialized cells called *melanocytes*. All people—white, black, albino—have the same number of melanocytes in their skin. The difference in color between one human being and another results from the amount of melanin produced in the melanocyte and stored in the skin.

Melanocytes are related to nerve cells. They start out in a part of the embryo called the neural crest and, before birth, move to the eyes, skin, hair bulbs, and brain. In human beings most melanocytes locate at the junction of the epidermis and dermis—the top and bottom layers of the skin.

Inside the melanocytes are an amino acid—*tyrosine*—and an enzyme—*tyrosinase*—which are needed to form melanin. Tyrosinase, like all enzymes, can catalyze or speed up a chemical reaction. In the presence of tyrosinase and a small amount of oxygen, tyrosine is converted to another amino acid, *dopa,* and subsequently to the black pigment, melanin.

When a mushroom or a potato is injured by cutting and exposed to air, plant tyrosinase accelerates formation of melanin. Tyrosinase in human melanocytes is similar to the enzyme in the mushroom and potato but not the

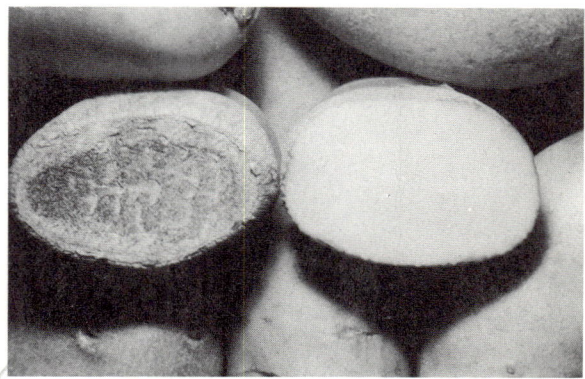

Fig. 2. A cut potato turns dark when exposed to air. The oxygen from the air combines with tyrosine to form melanin, and the process is catalyzed by the enzyme tyrosinase. A freshly cut potato is shown next to one with a darkened surface.

Fig. 3. Melanin pigmentation can be studied at different levels:

a. Looking at a sample of skin through a light microscope, one can see the pigment-producing cells, melanocytes, located at the junction of the epidermis and dermis. Pigment granules (melanosomes) are transferred from melanocytes to keratinocytes, the chief cells in the upper layers of the skin.

b. Focusing on the melanocyte through either a light or an electron microscope reveals the long tapering arms, dendrites, of the pigment cell.

c. The melanosome, a particle of pigment made in the cytoplasm of the melanocyte, is so small that it must be studied with an electron microscope.

d. Chemical conversion of tyrosine to melanin is catalyzed by the copper-containing enzyme tyrosinase located in the melanosomes.

same. A slight change in structure makes human tyrosinase different from that in other forms of life. All tyrosinases contain the metal copper.

Most people are born with normally working pigment cells. But in albinos tyrosinase does not function, so melanin cannot form. After the normal reaction between tyrosinase and tyrosine, the new melanin pigment fuses with protein to make a black granule. The melanin granule is incredibly small—about half a micron (μ), or one-millionth of a meter. When one looks at human skin that has been grown outside the body in tissue culture, one can see, with a microscope, melanin granules in the melanocytes and also being transferred to other cells in the skin.

We know from studying frogs and fishes that the colored granules in pigment cells move, collecting in the center or scattering throughout the cell. And as the granules move, the skin changes color.

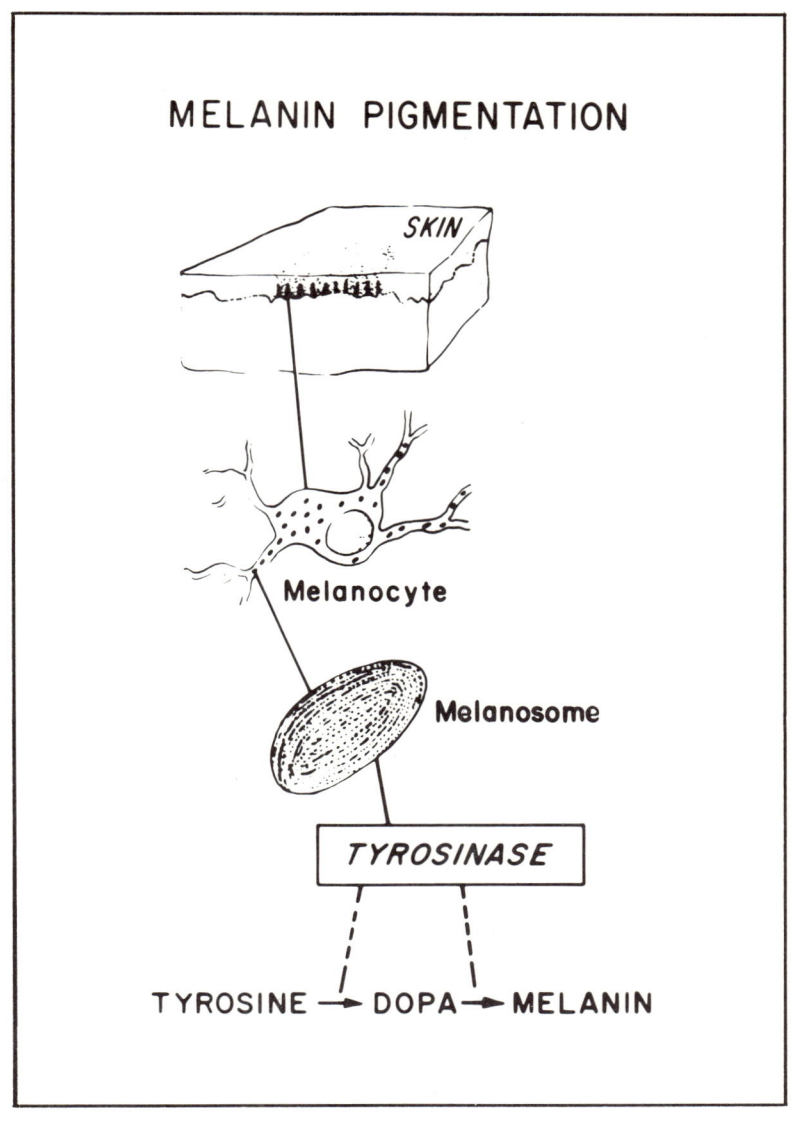

When the granules cluster in a tight little dot they are almost invisible, and the cell looks round and pale. The skin also looks pale because entering light is reflected from the white-appearing cells. But when the granules disperse, filling the body of the cell and its tapering branches, or *dendrites*, the cell is black and starlike. And the skin looks dark because entering light is absorbed by the black pigment and not reflected. The melanocyte does not contract or expand—only the pigment granules move. It is the redistribution of colored granules plus the absorption of light by the thin outer layers of skin that make possible a range of colors in fishes and frogs.

In man and all other animals more melanin means darker skin. Also, pigment dispersion contributes to color in frogs, fishes, and lizards. The more scattered the granules, the darker the skin. We do not know if pigment dispersion occurs in man and other mammals.

2

Color and Camouflage

Rapid changes in color are limited to a few groups of animals. Some animals without vertebrae, such as octopuses, squid, shrimps, and prawns, and others with vertebrae, such as fishes, frogs, and lizards, can change color rapidly. The capacity to change gives protection.

When a leopard frog, *Rana pipiens*, is placed on a dark background for several hours, it darkens. When the frog is placed on a white background with good illumination, it lightens. The frog's skin changes color as the melanin granules disperse—making the cells opaque, or clump together in a point—making the cells translucent.

The black spots that pepper the frog's skin are clusters of melanocytes. The green part of his uniform is created by brown and yellow pigments. When the melanocytes expose or withdraw their own colored granules, they cover or uncover other pigments in the frog's skin.

Fig. 5. *Right:* The frogs (*Rana pipiens*) and lizards (*Anolis carolinensis*) show the extremes of light and dark coloring. All animals were light in color one hour before being photographed. Injection of MSH (melanocyte stimulating hormone) into one frog and three lizards caused rapid darkening.

Fig. 4. *Below:* What happened is explained in the drawing of a microscopic view of lizard skin showing three different kinds of pigment cells—xanthophores, iridophores, and melanophores or melanocytes. The xanthophore acts like a yellow filter. Iridophores reflect light. Inside the iridophores are reflecting platelets—stacks of crystal-like plates that contain the silvery white pigment guanine. The dendrites of the underlying melanophores reach up and cover both the xanthophores and the iridophores. In the melanophore at the far left all the melanin granules are concentrated near the nucleus of the cell, leaving the dendrites free of melanin. The lizard's skin would appear bright green. After MSH is given, the pigment granules migrate into the dendrites of the melanophore, as shown in the middle cell, and the skin would be brown. Finally, in the cell at the far right, MSH has induced complete dispersion of pigment granules. The cell body is free of melanin, but the dendrites are full. The pigment within the xanthophores and iridophores is obscured by the melanin, and the lizard's skin would be dark brown.

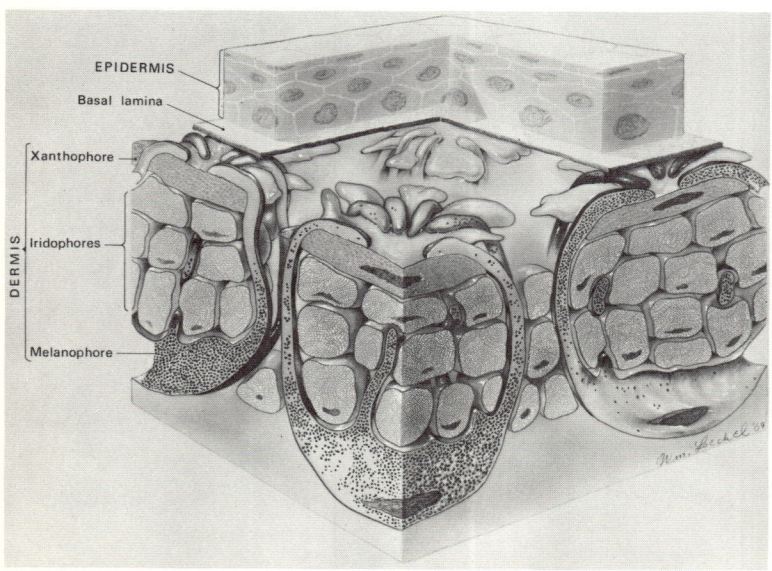

Fig. 6. *Above:* An albino bullfrog perches conspicuously beside a normal camouflaged relative. A yellow glint graces the albino's skin, for in the absence of melanin the yellow pigment in the xanthophores becomes visible. If the melanin were present in the melanocytes, and the xanthophores failed to produce their yellow pigment, the frog would appear blue, because the dark granules deep in the skin would absorb red to green light while the collagen above would reflect blue light.

Fig. 7. *Right:* From Australia—an albino wallaby (*Macropus rufogriseus*) with twin young in the pouch. Twinning is relatively rare in wallabies. The presence of two young in a pouch is not proof of multiple birth because sometimes a second young enters a pouch after dispossession by its true mother. However, the albino young in the photograph are not pseudo-twins sharing a pouch but genuine twins.

Fig. 8. Weasels (*Mustela erminea*) are shown in their brown summer and white winter coats. Weasels remain brown in southern regions, but they turn white in northern climates for camouflage.

Fig. 9. *Above right:* The hair of these mice was light brown a few months before their picture was taken. Hairs were plucked in a rectangular shape over the back of one animal and in the UC pattern, for University of California, in the second mouse. Both mice received daily injections of MSH during the period of hair regrowth with the result that the new hair grew out black.

Fig. 10. *Below right:* These litter mates were hairless at birth. After it was two days old, the mouse in the center received injections of MSH every other day for 20 days. Its hair grew out black. The mice that did not receive MSH developed the light brown hair seen ordinarily.

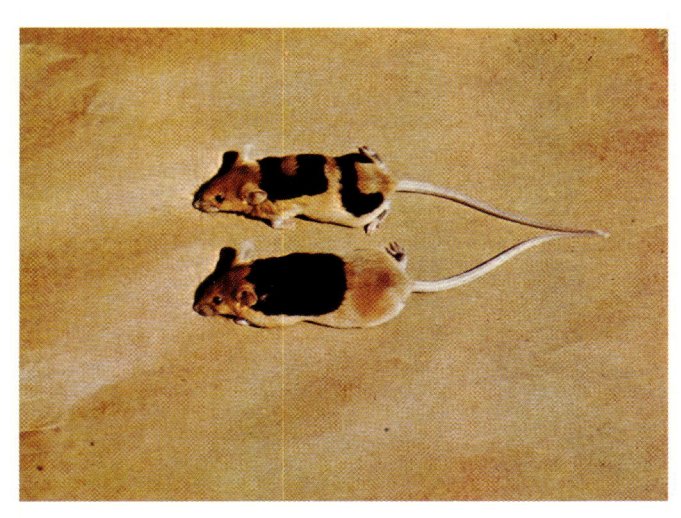

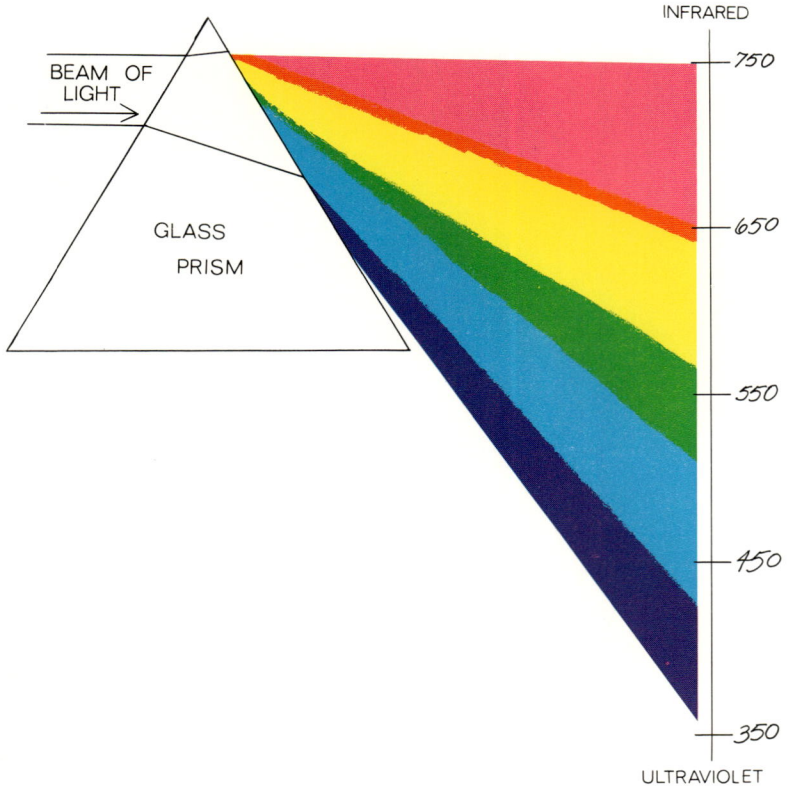

Fig. 11. White light from the sun, when passed through a glass prism, is divided into a rainbow of colors known as the solar spectrum. The wave lengths of visible light range from 350 nanometers at the blue-violet end to 750 nM at the red-orange end. Ultraviolet light from 290 nM to 320 nM, not visible to the human eye, is responsible for suntan and sunburn.

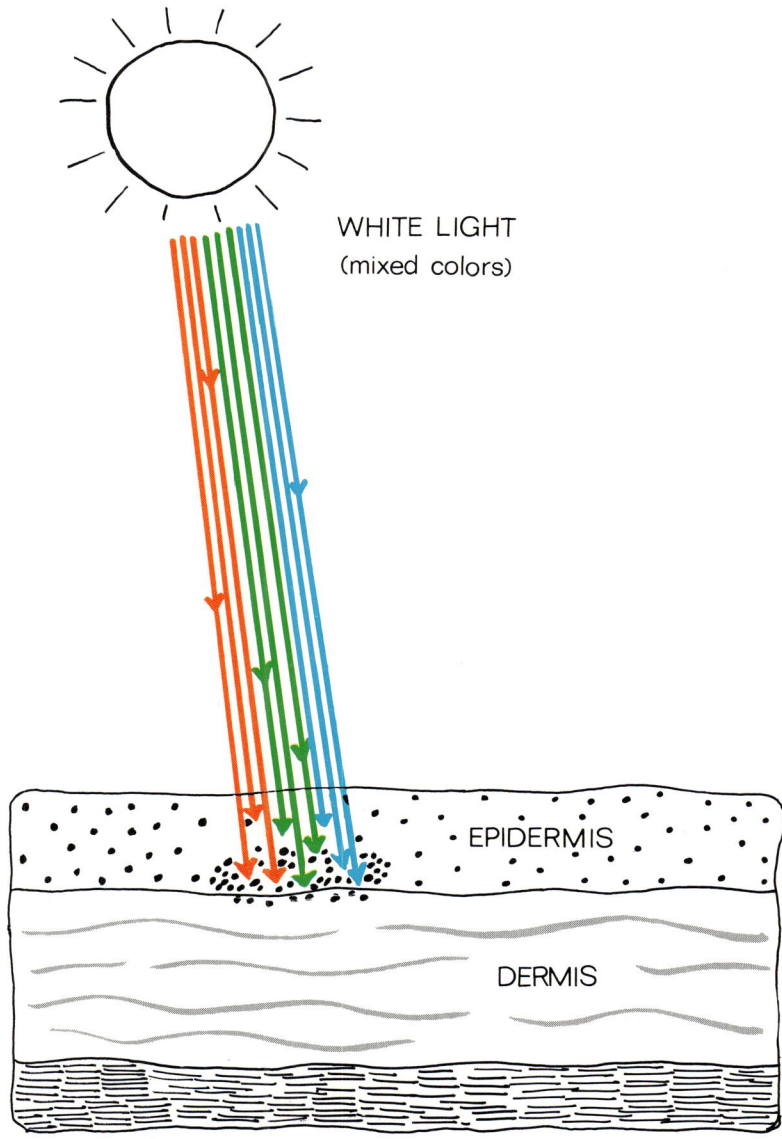

Fig. 12. When melanin is located within the top layer of skin (the epidermis) or at the surface of the bottom layer of skin (the dermis) the skin appears brown to black. Light of all wave lengths is absorbed by melanin and the skin.

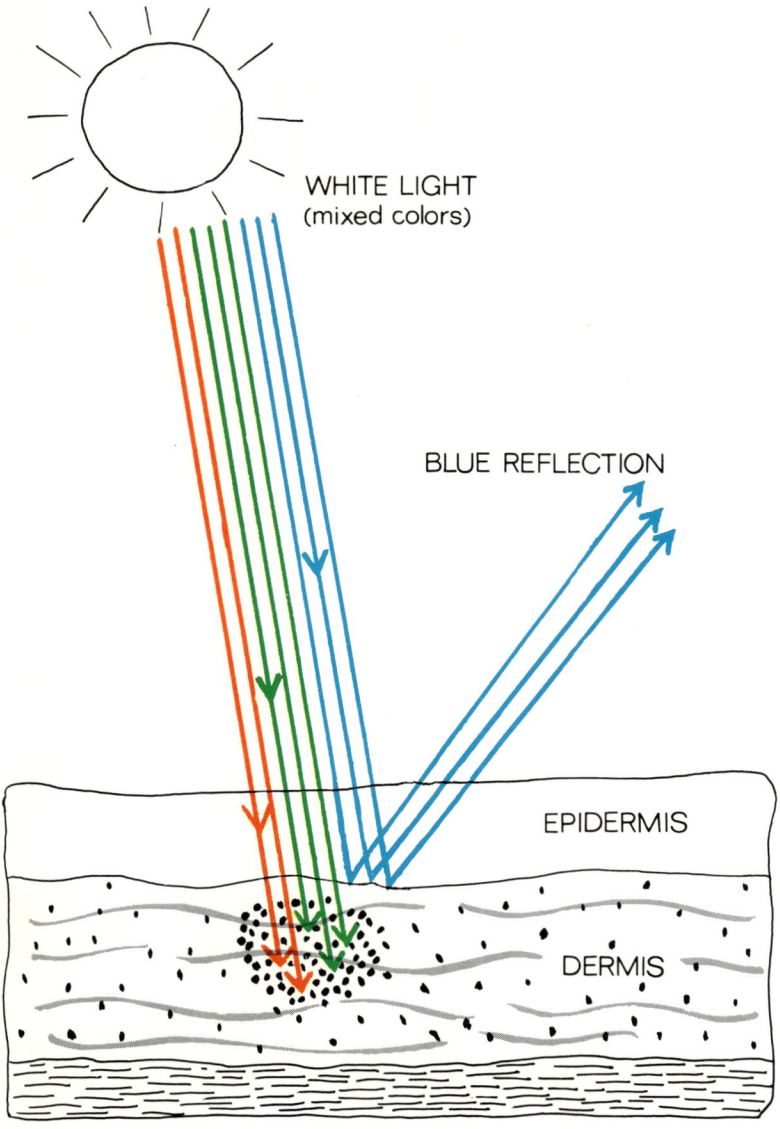

Fig. 13. Light of red to green wave lengths is absorbed by melanin, but blue light is reflected by collagen. Hence, dark pigment deep in the dermis appears blue.

Similar changes occur in fishes. Flounders and their flatfish relatives match their background skillfully. When put in a tank with a black and white checkerboard floor, they reproduce the pattern. But, like frogs, they need sight to play the game. A biologist kept gray catfish in a white tank until they bleached white. Then he blocked their eyes with goggles, and the fish darkened even though they remained in the white tank. Blinded fishes cannot match their background.

Two groups of killifish—one bleached in a white tank and the other darkened in a black tank—were allowed to swim together in a black tank. A penguin was added. At the end of the fish hunt, the count showed that 74 percent of the penguin's catch was bleached killifish. Some minnows are killifish. And fishermen know that minnows kept in a white container make good bait for pike and perch swimming in dark water.

Color changes are useful camouflage for reptiles as well as for frogs and fishes. Chameleons change color in minutes or seconds, especially when frightened. They also react to a different background, light, shade, or temperature.

The true chameleon, a native of Africa rarely seen in North America, shifts from its normal leafy green to brown, red, or black. The American chameleon—really a lizard named *Anolis carolinensis*—turns from bright green to brown in a few minutes. *Anolis* usually is green in sunlight and brown in shade.

Like fishes and frogs, chameleons change color as their pigment granules cluster or disperse. Under the surface of the chameleon's skin are particles of *xanthophyll*, the same pigment that makes the yellow of buttercups, dandelions, and autumn leaves. Below the xanthophyll layer are cells containing a silvery white pigment, *guanine*, that acts like a reflector. In the skin guanine appears blue rather than white. In bright light the blue color of guanine filters through the overlying yellow of xanthophyll, and the chameleon wears its usual coat of green. (Guanine crystals obtained from fish scales are used in cosmetics. The frosted look of nail polish and lipstick comes from guanine.)

At a deeper level in the skin of the chameleon are the melanocytes with dendrites branching towards the surface. When the melanin granules spread upwards into the dendrites of the melanocytes, hiding the particles of guanine but not xanthophyll, the chameleon's skin turns brown. When the melanin covers xanthophyll as well as guanine, the animal looks much darker.

Some creatures fit neatly into their surroundings without changing color. The green color of animals living among plants and grass—insects, frogs, snakes, lizards, and tropical birds—is produced in different ways. Some green caterpillars have green blood because of a pigment derived from the chlorophyll in their plant food. Stick insects, grasshoppers, and many caterpillars also make a green pigment, but it is not the same as chlorophyll.

The skin of birds and mammals, covered with feathers and fur, does not change color rapidly. A few mammals change their hairy coats with the seasons. In North America the prairie hare and several other varying hares, including the snowshoe rabbit, turn white in winter. However, the cottontail does not change color.

In weasels the seasonal change varies with the climate and latitude. The ermine, *Mustela erminea*, and the dwarf or least weasel, *Mustela rixosa*, turn white in winter in the Soviet Union, northern Scotland, Canada, and in some northern areas of the United States. Their white winter coat provides the ermine used for royal robes. The mink, a relative of the weasel, does not change color, although there is a white, black-eyed variant that remains white throughout the year. The polar bear and the snowy owl, which live in arctic regions, also wear white all year around. Many animals seem naturally painted with no obvious relation to camouflage, for example, the reindeer and the raven.

Seasonal change in the length of the day and in the intensity of daylight affects hair growth and shedding in the weasel and the mink, and in some other animals. The amount of daylight also affects the feather pigments of some birds but has no obvious influence on color in mammals.

3

Black, White, Red, Yellow

In animals that change color rapidly, such as frogs, fishes, and lizards, the pigment granules move in response to differences in background color and light. What makes the granules move?

Information about the environment reaches the brain through the eyes and is carried to the pigment cells by hormones or nerves. Also, light and temperature can act directly on the melanocytes in the skin.

Frogs and fishes react to a substance called melanocyte stimulating hormone or MSH. Light enters the frog's eyes and excites a structure in the brain to release MSH. The hormone, carried by the blood to the pigment cells, stimulates the colored granules to move. If a small amount of MSH is injected into a frog, the pigment granules disperse throughout the melanocytes, and the skin darkens rapidly.

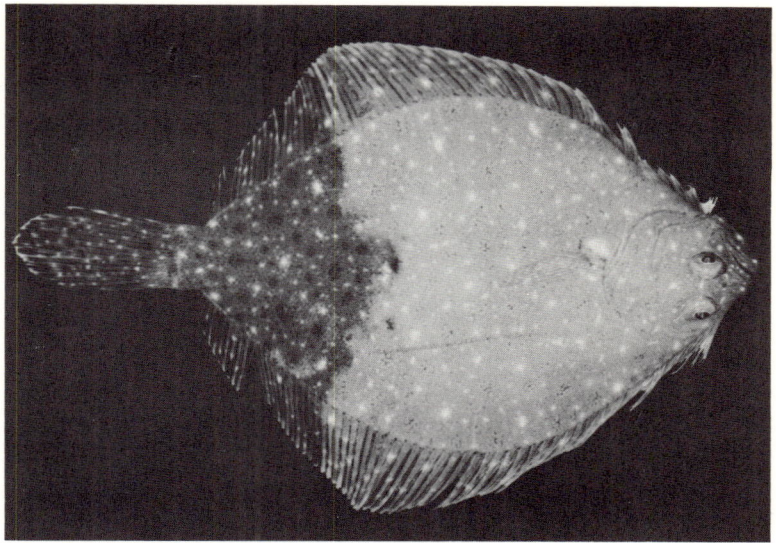

Fig. 14. A sand flounder (*Scopthalmus aquosus*) shows darkening of the lower part of the body and fins. The darkening occurs after the nerves to that part of the body have been cut.

When MSH is injected into people with normal pigmentation, their skin darkens, but more slowly than the frog's. We do not know whether melanin granules move inside human melanocytes as they do in the frog's. The darkening effect of MSH can be reversed in frogs by another hormone, melatonin. But melatonin has no effect on skin color in man.

If the nerve to the tail fin of a fish is stimulated with an electric current, the fin lightens. If the nerve is cut, the fin darkens. Perhaps a substance normally released at the nerve endings keeps the fin light. If so, after the nerve is cut the lightening factor is lost, making the fin darken.

The pigment cells in frogs react differently from those in fishes. When the nerve to a frog's leg is cut, the skin does not darken or lighten. The effects of hormones and nerves on pigment cells differ in different species.

In man melanin is the chief coloring factor in skin, hair, and eyes. Other pigments add or subtract various hues.

Hemoglobin is the pigment in oxygenated blood that makes skin look pink—or red when a person blushes. The pinkness from circulating blood is most visible through thin, untanned skin containing little melanin. If the blood does not get enough oxygen, the hemoglobin is more blue than red, and the skin looks blue. Too little blood circulating near the surface of the body, when a person faints or bleeds severely after an accident, makes the skin look pale.

Sick people who have excess water in their skin also look pale. If hands or feet are bandaged tightly, or allowed to stay wet for a long time, the skin becomes soggy, wrinkled, and whiter than usual. Albinos look white or pink because they cannot make melanin, which lends a tan hue to skin and partially hides the color of circulating blood.

Carotene is a yellow pigment present in the dead outer layers of skin and in body fat. Where skin is thick, as in the palms and soles, and where fat is abundant, the yellow of carotene is visible. Women store more carotene than men, but carotene plays no part in differences among races.

Carotene makes the yellow of butter and carrots but, compared with melanin, adds little to the skin color of human beings. If a person eats enough carrots to overload his blood with carotene, then his skin looks yellow. Healthy people, no matter where they come from, do not have yellow skin—any more than the North American Indian has red skin.

4

Suntan-Sunburn

When sunlight is passed through a glass prism, it is broken up into a dazzle of colors known as the spectrum. The series starts with red and goes on to orange, yellow, green, blue, and violet. The wave lengths making up the solar spectrum range from 750 nanometers (nM) at the red end to 350 nM at the violet. (A nanometer is one-billionth of a meter.) Ten percent of all the energy from the sun comes from the ultraviolet end of the spectrum — the region beyond the violet that consists of very short wave lengths measuring 480 nM or less. Sunlight increases skin color. And the small portion of the sun's spectrum between 320 nM and 290 nM is the complete suntan-sunburn range.

Getting a sunburn or a suntan depends on how susceptible an individual is to ultraviolet light and how intense the light is. The darker the person, the less likely he is to get a sunburn, because the melanin in his skin acts like a shield and cuts down entry of damaging ultraviolet rays. The horny layer at the skin's surface also protects the skin against ultraviolet radiation.

The intensity of sunlight depends upon the distance between sun and earth, the thickness of the earth's atmosphere, and the height of the sun above the horizon. As light passes through the atmosphere, it may be absorbed or scattered back into space by tiny particles in the atmosphere. The scattered light contains more short wave lengths—those from the blue-violet end of the spectrum—than long wave lengths—those from the visible and infrared regions. Hence, when clouds, fog, dust, and smoke thicken the atmosphere, less ultraviolet light gets through to earth.

It is easy to sunburn on a mountain because more ultraviolet light penetrates the thin atmosphere, and in addition, a lot of light is reflected from snowcaps. Sand also reflects ultraviolet light, so sunburning is common at the beach. It is hard to tan early in the morning or late in the afternoon, even on a bright summer day, because when the sun is low in the sky its oblique rays must pass through more atmosphere.

How does ultraviolet light from the sun or a lamp cause burning? On intense exposure to ultraviolet light the skin gets sunburn red because small blood vessels near the surface of the skin enlarge, their blood flow increases, and skin temperature rises. The epidermal cells are injured, and the skin looks and feels burned. In a few days the damaged cells peel off, and other cells divide rapidly to replace those that are shed. The stepped-up multiplication of cells more than doubles the thickness of the epidermis, and it remains thick for a few weeks.

A person can tan without burning. Immediately after exposure to ultraviolet light, pigment that was already in the melanocytes, but in a bleached form, begins to darken. But the early tanning is lost within 24 hours. Three or four days after a person exposes himself to ultraviolet light, darkening develops that lasts for several weeks. This slower, persistent tan comes from new pigment formed when tyrosinase in the melanocytes is activated by ultraviolet light.

Even without sunburning, repeated heavy exposure to ultraviolet light for many years produces thickening of the skin, brown marks that differ from ordinary freckles, and skin cancer.

A blue-eyed, fair-skinned person with blond or red hair is more susceptible to the aging effects of sunlight than a person with darker coloring. Leathery skin, etched with lines and spatterdashed with brown spots and horny bumps, is the mark of the outdoor person. The farmer, sailor, fisherman, and golfer—people who live a lot outside—have the greatest chance of being exposed to sunlight and developing its long-term effects.

5

Skin Spots

Black, brown, red, white, and blue—the colors of spots in human skin. Everybody old enough to look can find a mole or a freckle on his body. Less common are white spots, blood marks, and tattoos.

Very few people, less than 3 percent, are born with moles or birthmarks. A mole, also called a *nevus*—or nevi for more than one—usually appears after age three or four. Pigmented nevi contain clusters of melanocytes. In the small, flat, dark mole, pigment cells nest at the junction of the two skin layers, the epidermis and dermis. This common, hairless mole is called a junction nevus. Junction nevi appear throughout life, especially after an illness, or during adolescence and pregnancy.

The raised mole, with or without hair, consists of pigment cells located in the dermis. It is called a dermal nevus.

The word nevus also is used to describe spots made up of small blood vessels or capillaries. The blood-filled nevus is salmon, wine, or flame colored and does not have pigment cells.

Dark spots seen after age 40 on exposed areas like the face and backs of the hands are mistakenly called liver spots. They have nothing to do with the liver, and their real name is *lentigines*—or lentigo for only one. They are darker than freckles and, unlike freckles, they occur singly.

Freckles are different from nevi and lentigines. They begin in children about five years old after they have been out in the sun. Freckles become darker and more numerous in summer than in winter. The face, neck, and backs of the hands are the common sites for freckles. They may fade completely or just lighten in color as a person grows older. Freckling is inherited, and the trait usually is linked with red hair.

White spots that are present at birth may point to partial albinism or to a rare neurologic defect. White spots that appear after birth often result from loss of the ability to form pigment, a disorder called *vitiligo*.

Scars from burns, chemicals, cuts, or other wounds are white if the melanocytes were destroyed when the skin was injured.

Clusters of pigment cells deep in the dermis form a blue nevus or, when spread out, a Mongolian spot. The blue color results from subtractive color mixing. The bulk of the protein in skin is collagen. In the lower part of the dermis collagen reflects blue light while the melanin within the dermis strongly absorbs light between the green and red. Thus the blue reflection is isolated by removal of light of longer wave lengths.

Not all blue, black, or brown spots are melanin. The black-and-blue stain of a bruise comes from broken red blood cells. Tattoos are made of nonmelanin pigments buried in the skin accidentally or on purpose. Spattered pigmentation can follow an automobile crash if dirt penetrates unbroken skin at high speed. Coal miners get blue gray stripes if they cut or scrape their skin and coal dust fills the wound.

The amateur tattooist pushes India ink or soot into his skin with a pointed object. The professional uses an electric needle to inject pigments deeply. Tattoos get their light blue tint from cobalt, dark blue or black from carbon, red from mercury, green from chromium, yellow from cadmium, and brown from ocher or iron ore. Tattoos can be troublesome. People who get tattooed usually are in their teens or early twenties. Later almost all of them wish they did not have tattoos. Although some pigment particles push out with time, most must be cut out. Such an operation substitues a small scar for a large tattoo.

6

Skin, Hair, and Eyes

People—black, white, and albino—have the same number of pigment cells in their skin—about 1,500 per square millimeter of skin surface. The total number of melanocytes in an adult's epidermis is about two billion. Given equal numbers of melanocytes, why the difference in skin color?

Albinos, whether of black or white parents, are completely white at birth. Born without tyrosinase, the enzyme required for the oxidation of the amino acid tyrosine, they cannot make melanin. The cause of partial albinism, in which localized white spots appear at birth, is unknown. Albinism is a rare disorder and does not account for normal differences in skin color.

The amount of pigment in the skin depends chiefly upon man's inherited capacity to produce melanin. Genes, the hereditary units made up of DNA—*Deoxyribose*

Nucleic Acid—determine how much melanin a person has and how fast it is made and lost. After the pigment is formed in the melanocytes, it may stay there or be transferred to other cells in the skin, thus contributing to the color of the individual. Melanin can be lost in two ways. Pigment granules leave the skin through normal shedding of the top layer. They also are absorbed by the body's tissues.

More is known about inheritance of skin color in mice than in men. A mosaic mouse, striped black and white from its back to its belly, was born in a laboratory. The mouse was the offspring of four parents and had inherited different colors from its two mothers and two fathers. How did such a hybrid form? A biologist mated two pure black mice to yield fertilized eggs and two white mice for their eggs. Normally, after an egg cell is fertilized it splits into 2 cells, then 4, 8, 16, and so on, to form a multicellular embryo. When the fertilized eggs from the black female and the white female had divided into 8-cell groups, the biologist removed them from the two mothers, put them into a growth medium, dissolved the membrane covering each embryo, and pushed together the egg groups from each mother. About 24 hours later the single combined embryo was transplanted into the uterus of another female mouse. The baby that grew in the incubator mother was born with black and white stripes.

Some laboratory-bred mice develop a patchwork of colors, but others are all black or all white. Growing mosaic mice helps to advance knowledge about color in mammals.

Fig. 15. Mosaic mice were developed by combining paired embryos obtained from two separate pairs of parents of different genetic backgrounds. By mixing developing eggs removed from a white donor female (that had mated with a white male) with eggs removed from a black donor female (that had mated with a black male) and transplanting the single combined embryo into another female mouse that served as an incubator, offspring with artificial coat colors were produced. A standard pattern of pigmentation, as shown in the diagram, consisted of broad transverse stripes of alternating colors extending from head to toe.

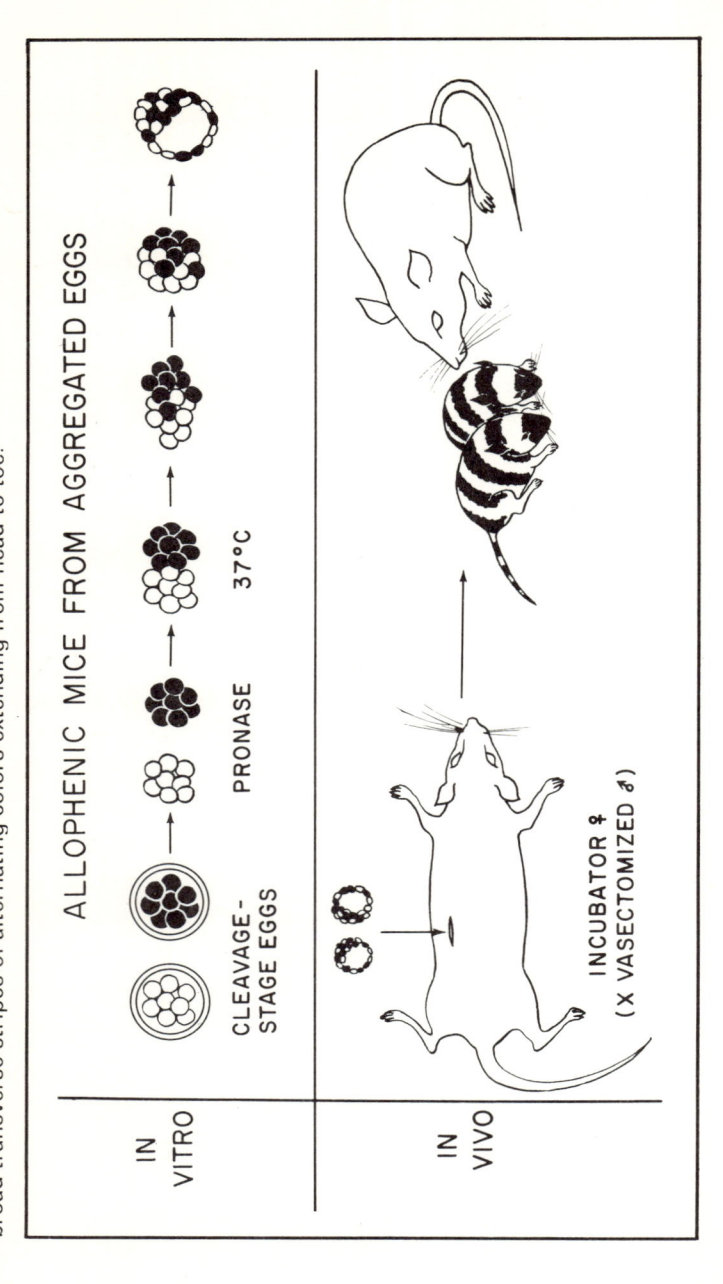

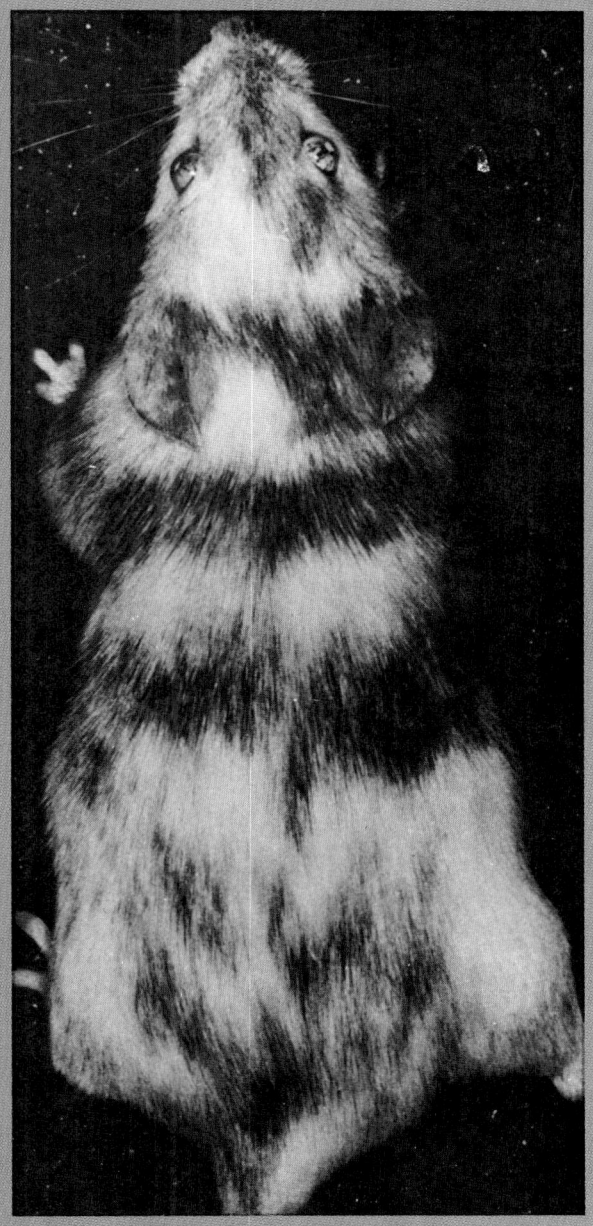

Fig. 16. An adult mosaic mouse shows alternating black and white transverse stripes.

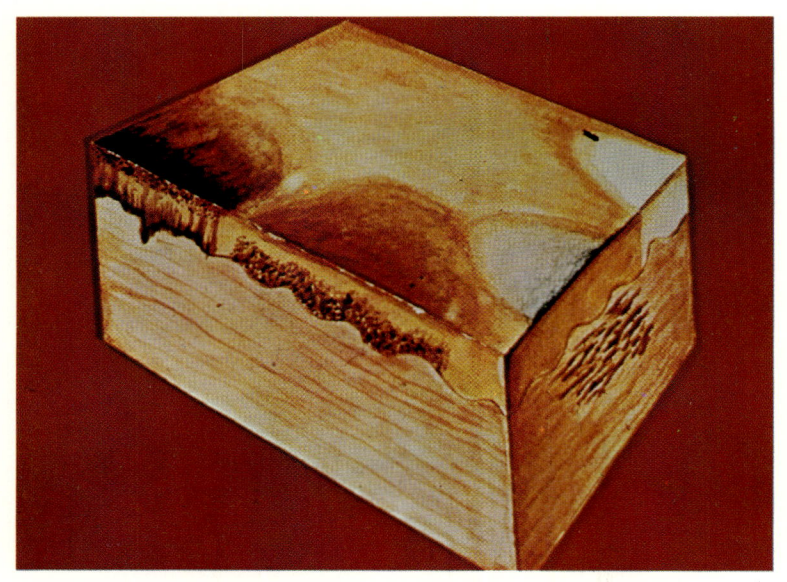

Fig. 17. This three-dimensional drawing of a section of skin shows four kinds of pigment changes. From left to right: (1) a pigmented nevus (mole) resulting from an increase in the number of melanocytes at the epidermal-dermal junction as well as an increase in the amount of melanin; (2) a freckle caused by an excess of melanin but normal numbers of pigment cells; (3) a blue nevus, the result of an increase in both melanocytes and melanin deep in the skin; and (4) a patch of vitiligo, in which melanocytes are lacking.

Fig. 18. *Above right:* A professional decorative tattoo, in which red and blue pigments have been injected deep into the skin.

Fig. 19. *Below right:* This unusual photograph shows that coal miners sometimes receive tattoos when their skin is injured and coal dust enters the wound.

43

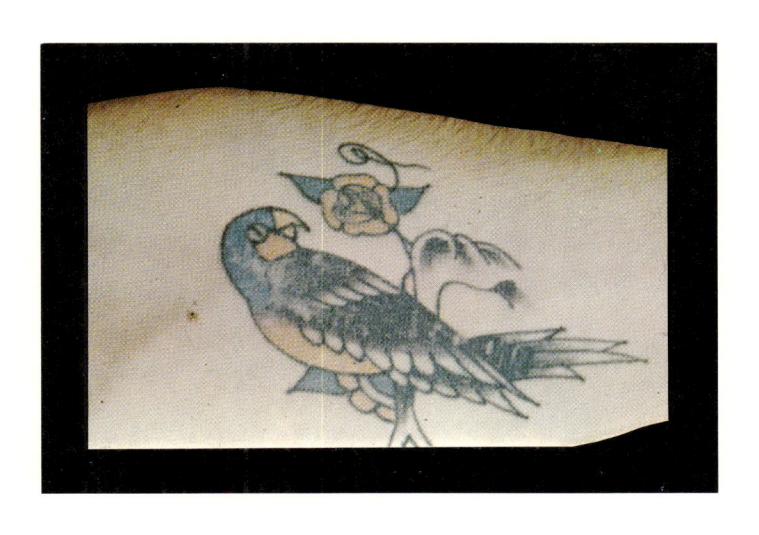

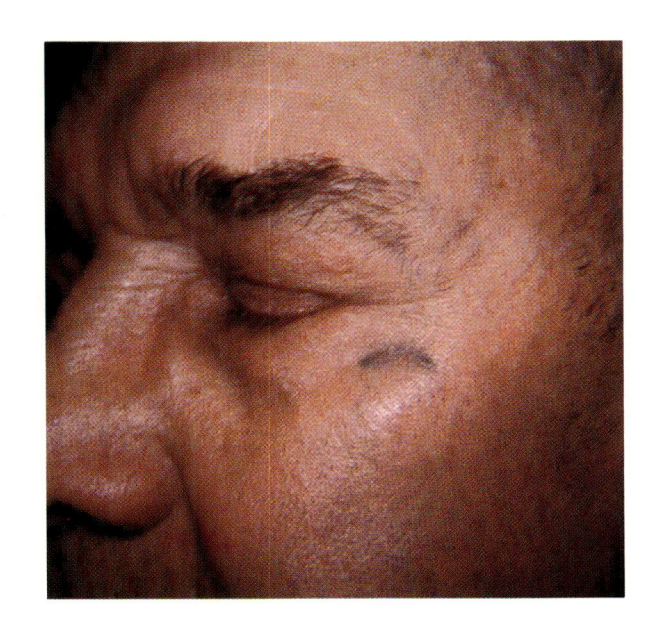

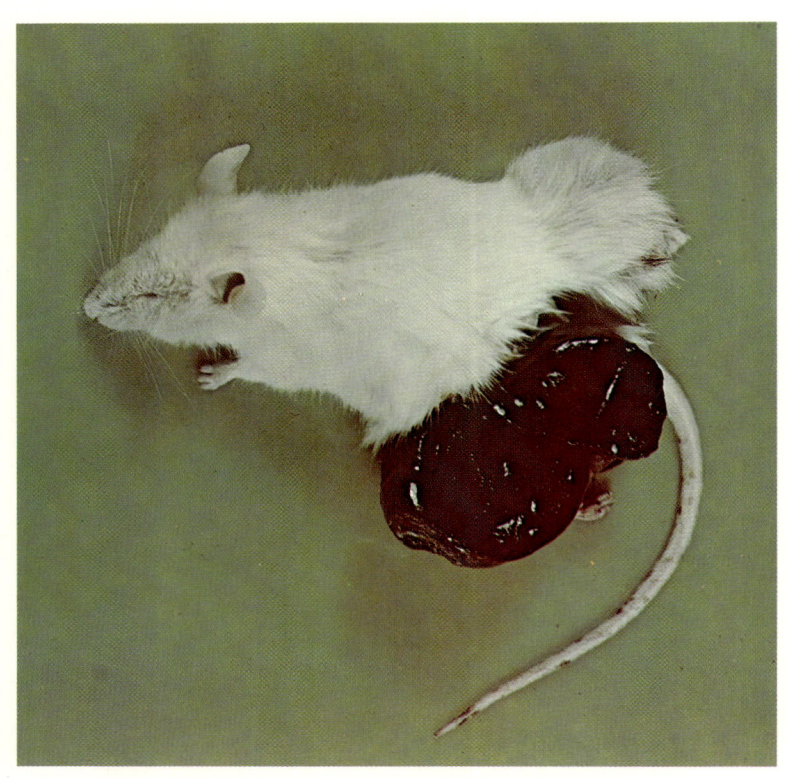

Fig. 20. An albino mouse with a large black melanoma. A piece of a tumor (known as the Harding-Passey mouse melanoma) was injected into the left and right hind legs of the mouse. Two months later the tumors were almost as large as the animal. For this photograph the skin was removed from one of the tumor masses but left intact over the other.

Fig. 21. The carrot-topped boy and the golden retriever are 14 years old. The boy's hair and the dog's fur match the reddish-brown rabbit, guinea pig, and mouse. Red pigment is formed differently from other natural colored substances. The first product of the tyrosine-tyrosinase reaction on the way to melanin is the amino acid dopa. If dopa combines with a different amino acid, cysteine, before continuing the steps leading to melanin, a red substance, rather than a black one, results.

Fig. 22. The five-year-old girl, with skin as fair as Snow-White's, is an albino. The rabbit, guinea pig, and mouse also are albinos. Lack of pigment in her eyes makes the child so sensitive to bright light that she squints when out in the sun.

Fig. 23. The 13-year-old boy has black skin, hair, and eyes. The rabbit, mouse, and two-year-old Labrador retriever are black, but the guinea pig is black and brown. The boy has the same number of melanocytes in any given area of his skin as the red-headed boy and the girl with albinism. Except for redheads, all normally pigmented people—no matter what their geographic, anatomic, or social classification—are colored with melanin, the same in kind but different in amount.

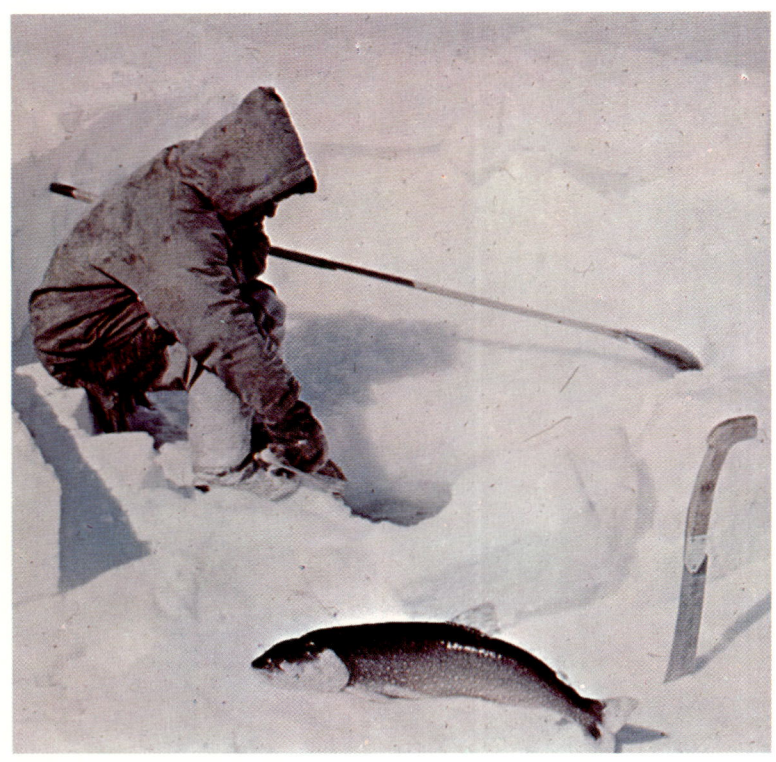

Fig. 24. An Eskimo fishes for char in the arctic regions of Canada. He, like other Eskimos, receives vitamin D, which is essential to human life, from fish liver instead of sunlight.

In the mouse at least 25 genes control melanin formation in the skin, coat, and eyes. The size, shape, and degree of darkening of pigment granules vary in different coat colors.

Inheritance of skin color in man involves more than one gene, but we do not know how many. Melanin granules tend to be more numerous and larger in dark skin than in light skin. It is obvious that parents with black skin produce children with black skin, and light-skinned children come from light-skinned parents. A child can be darker or lighter than either parent, but the differences are small.

When a black person marries a white person, the child's skin color is between that of the parents. And the child carries hereditary factors for black and white. If each parent is the child of a black and white marriage, their children can be black, white, or between the extremes.

Hair ranging from black to blond (but not including red) contains the same pigment. However, the lighter the hair color, the more dilute the melanin. In blond hair the melanin granules are less numerous and smaller than in black hair. The pigment in black and blond hair is made only from tyrosine, but the pigment of red hair is formed by a combination of a tyrosine product with another amino acid, cysteine.

Eye color depends upon the amount of pigment in the iris—the colored muscle surrounding the pupil—and on optical effects. When light enters the pupil, some of it is absorbed by the pigment granules in the iris, and the rest is reflected. Brown or black eyes have a lot of melanin—thus little light is reflected. Blue eyes do not have blue pigment; they have fewer melanin granules, so that less light is absorbed. Pink eyes occur in albinos, who do not have enough pigment to hide the blood vessels in the iris. All light entering the pupil is absorbed by the blood vessels except red light, which is reflected back from the blood.

7

Why Melanin?

People from the warm regions of our planet have more melanin than those from the cool areas. From Africa at the equator, northward to southern Europe, northern Europe, and Scandinavia, human skin color changes from black to near white. The color of eyes and hair tends to go along with skin color. Nearly all Africans, Indians, and Orientals have dark eyes and hair. Why are people living near the equator darker than those from cold areas?

One anthropologist believes that in the African jungles man needs dark color for camouflage. However, his idea cannot be extended to account for the tapering off in melanin pigmentation as one goes north. Another explanation is that lightly pigmented men could not survive cancers induced by sunlight in the tropics. But this one-sided theory does not explain why fewer dark people live in the cold regions.

A different view is based on the experimental evidence that vitamin D is necessary for life. Without vitamin D growing children develop rickets, a bone disease that can be crippling. The body can get vitamin D from only two sources: food, and the action of sunlight on skin. A precise amount of vitamin D is required. Too much produces toxic effects that are as bad as too little.

Light skin is efficient in making vitamin D after exposure to sunlight. Dark skin is inefficient because its melanin absorbs most of the light energy. Light-skinned people could survive in the colder regions because they could make a lot of vitamin D even with little sunlight. Dark-skinned people could exist in the warm regions because they avoided an overproduction of vitamin D. The only dark-skinned people who could live in the cold were Eskimos, who regularly ate fish liver containing vitamin D. We do not need skin-made vitamin D today because enough occurs in food. Interesting as this concept is, it still cannot be accepted as the full explanation for variations in the color of people in different climates.

What then is the value of melanin? There are no disadvantages, from a medical standpoint, in having a lot of melanin. The big advantage of the pigment is that it supplies a built-in shield against sunlight. People with little melanin suffer skin damage leading to increased wrinkling and cancer. The darker the skin, the less chance of developing skin cancer.

Frogs, tadpoles, fishes, and chameleons change color in seconds or minutes. Can man do the same? Do not say no immediately. Some people believe that the skin below their eyes varies from light to dark depending upon how they feel. A person may have dark shadows when he is tired and light skin beneath his eyes when he is rested. Perhaps this change in color results from variations in melanocyte activity regulated by the nervous system.

The skin becomes excessively dark or light in some diseases. Hormones produced abnormally can make a patient dark, but after proper treatment the skin lightens to its normal color. Patients with a melanoma, a black cancer starting from pigment cells, may also become dark.

Among the people who have vitiligo, a few lose all the pigment in their skin after birth and become totally white. They look like albinos except that they keep their original eye color, and their hair may or may not turn gray.

A young man once asked, "If all the melanin were taken out of the skin of a black man, would he be a white man? And if that melanin were put into a white man, would he be black?" People change their names. They dye their hair. Some day they may be able to regulate their skin color and become light or dark at will.

ACKNOWLEDGMENTS

The illustrations are reproduced through the courtesy of:

Fig. 1. American Museum of Natural History

Fig. 2. Dr. Eugene Mirrer

Fig. 3. Drs. J. Duchon, T. B. Fitzpatrick, and M. Seiji. The chart is from their paper, "Melanin 1968—Some Definitions and Problems," in *Year Book of Dermatology,* 1967-1968 Series.

Fig. 4. Dr. Mac E. Hadley. The chart will appear in a paper by him and Dr. J. D. Taylor, "Chromatophores and Color Change in the Lizard *Anolis carolinensis,*" Z. Zellforsch, vol. 104, 1970.

Fig. 5. Dr. Sidney Hurwitz, who photographed the lizards and frogs with a Nikon F camera and a micro Nikkor 55 mm lens and Kodachrome II film. The albino bullfrog was raised from a tadpole by Charles Miller, a Yale College student.

Fig. 6. Dr. Richard B. Swint, who used a Hasselblad 500 C camera with an 80 mm lens and high speed Ektachrome film. Assistants were Margaret D. Koonce and John R. Hendee.

Fig. 7. Dr. E. M. Nicholls, who obtained the photograph from the son of the late Sir Edward Hallstrom. The photograph was taken by Sir Edward, who kept a colony of albino kangaroos in Sydney, Australia.

Fig. 8. Drs. Charles C. Rust and Roland K. Meyer

Figs. 9 and 10. Dr. George T. Scott

Figs. 11, 12, and 13. Independent Picture Service

Fig. 14. Dr. Irving I. Geschwind

Figs. 15 and 16. Dr. Beatrice Mintz, who used the chart in her paper, "Gene Control of Mammalian Pigmentary Differentiation, I Clonal Origin of Melanocytes," from the *Proceedings of the National Academy of Sciences,* vol. 58, July 1967.

Fig. 17. Drs. T. B. Fitzpatrick, K. A. Arndt, W. H. Clark, Jr., A. Z. Eisen, E. Van Scott, and J. C. Waughan. The drawing is from their book, *Dermatology in General Medicine,* New York: McGraw Hill, 1970.

Fig. 18. Dr. Herman Beerman

Fig. 19. Dr. William A. Welton

Fig. 20. Independent Picture Service

Figs. 21, 22, and 23. Stanley Schulz (redhead), Donna Verry (albino), and Leon Horton (black), the subjects in these photographs. The Labrador retriever belongs to Turan E. Onat, and the golden retriever is the canine member of the author's family.

Fig. 24. National Film Board of Canada

Index

Numbers in italics refer to illustrations.

albinos, 10, 12, 30, 38, 44, *46*, 50
amino acid, 11, *45*, 49
atmosphere of the earth, 33

birthmarks, 35
black skin, 10, *47*, 49, 51, 53

camouflage, 16, *18*, 25, 27
carotene, 30-31
chlorophyll, 26-27
collagen, *24*, 36-37
cysteine, *45*, 49

dendrites, 14, *16*
dermis, 11, 35-36
dispersion, of pigment, 14, 15, *16-17*, 26, 28
DNA, 38-39
dopa, 11, *12-13*, 45

enzymes, 11-12
epidermis, 11, 34, 35, 38
ermine, *20*, 27
eyes, 9, 11, 30, 50, 51

feathers, 9, 27
freckles, 34, 35, 36, *42*
fur, 9, 27

genes, 38, 49
guanine, *16*, 26

hair, 9, 11, 30, 45, 47, 49, 51
hemoglobin, 30
hormones, 28-29, 53

lentigines, 36
leopard frog (*Rana pipiens*), 15, *16-17*, *18*
liver spots, 36
lizard (*Anolis carolinensis*), *16-17*, 25-26

melanin: in dinosaurs, 9; in fishes, 12, 14, 15, 25, 28-29; in frogs, 12, 14, 15, *16-17*, 28-30; in the lizard, *16-17*, 25-26; in mammals, 10, 14; in man, 10, 29, 34-39, 47, 49-53
melanocytes: definition of, 10, *13*; dispersion in, 12, 14, 15, *16*, 26, 28-29
melanocyte stimulating hormone (MSH), *16-17*, 28-29
melanoma, 12, *44*, 53
melanosomes, 12, 13
melatonin, 29
mice, 20, *21*, 39, *40-41*, 49
mole (skin spot), 35, 42
Mongolian spot, 36

nevus, 35-36, *42*

plant pigmentation, 9-10, 11

red skin, 31

scars, 36
skin cancer, 34, 51-52
solar spectrum, *22*, 32
sunburn, 22, 32-34
suntan, 22, 32-34

tattoos, 35, 37, 42-43
tyrosinase, 11-12, *13*, 34, 38, *45*
tyrosine, 11-12, *13*, *45*, 49

ultraviolet light, *22*, 32-33, 34

vitamin D, 52
vitiligo, 36, *42*, 53

weasel, *20*, 27
white skin, 10, 49, 52-53

xanthophyll, 26

yellow skin, 30-31

The Author

Marguerite Rush Lerner, MD, is an associate clinical professor of dermatology at Yale University School of Medicine. Among the books she has written are *Where Do You Come From?*; *Red Man, White Man, African Chief*; and *Who Do You Think You are?*. Dr. Lerner's first academic degree was earned in English literature at the University of Minnesota. She attended Barnard College for pre-medical work and studied medicine at Johns Hopkins and Western Reserve universities. Dr. Lerner lives with her husband and four children in Connecticut.